The Path to
The Military Reception Center

Vickey Joins the Military Series

Book 1

The Path to
The Military Reception Center

Book One

Table of Contents

High School Graduation

The big day had come—it was graduation time. Everyone was happy and a little scared about what comes next.

When the celebration was over, it was time to think about the future.

Looking in the mirror, I was proud of what I saw: a girl who admitted she was scared but had something she was determined to accomplish.

As the time had come to
move on, I, the class
joker, hoped my jokes
and laughter would help
me on my new adventure.

Truth be told, I wasn't completely sure if I was ready to go out and make my mark in the world, but I was determined to give it everything I had. I was chasing success—not just in my career but in shaping who I was as a person. Whatever path I chose, I wanted to excel at it.

High school had been my safety net, a familiar place where I knew every hallway and every face. Leaving it behind felt like removing the training wheels on a bike—exciting but terrifying. Despite my fears, I was as prepared as I could be to take the leap into adulthood.

I was SCARED, but I was ready to try it; at least as ready as I was ever going to be. The idea of stepping into the unknown, making decisions that could shape my entire future, was overwhelming. But ready or not, I knew I had to try. It would be my time to see what I was truly capable of.

Decision Time

Suddenly, I was no longer a high school student, no longer blanketed by the predictable and structured world of school—the place that outlined my every waking moment. I was no longer nestled in a secure environment where the outcome was clear and the road to achieving that outcome defined.

Now, the future was mine to shape; no one else would make decisions for me or dictate my thoughts. I was a graduate, standing at a crossroad, ready to choose my path. **It was decision time**.

For much of my life, I had dreams and aspirations of being a journalist. I dreamed of following in the grand footsteps of Oprah Winfrey and Barbara Walters.

So, what was the next step? Believing that a successful career in journalism started with a solid education, I applied to university. When the acceptance letter arrived, it confirmed the start of a new chapter. It would be a journey for which I was sorely unprepared.

But wait………. there was something else that tugged at me; something else I really wanted to do: joining the military.

I actively pursued the option. I visited the recruitment center, took the entrance exam, and even searched for a writing job in the military. However, such a role came with no guarantees.

Unable to fulfill my two dreams, journalism and the military at the same time, saddened me, but the decision was easy, journalism. It was not a difficult decision for me because being a journalist was a stronger, longer-desired dream.

So, I made the decision. I left the recruitment center and went to prepare for my future as a journalist. Scared, but excited, a few months later I enrolled as a first-year student in the School of Journalism.

The College Years

The journey started off on a rocky note. The first year was tough, to say the least. None of my friends had joined me on my college journey, so I was alone, and the first year was full of uncertainties, bewilderment, and sadness.

But through all the awful emotions, one thing remained steady: continuing the lifelong dream of obtaining a journalism degree so that I could follow my passion for writing.

Socially, things began to come together.

By the time the second year rolled around, I needed to put an end to the loneliness. Therefore, I set out on a journey to find friends. However, the campus and university life were daunting, so I began my journey at the student center—a hub of activity and potential friendships.

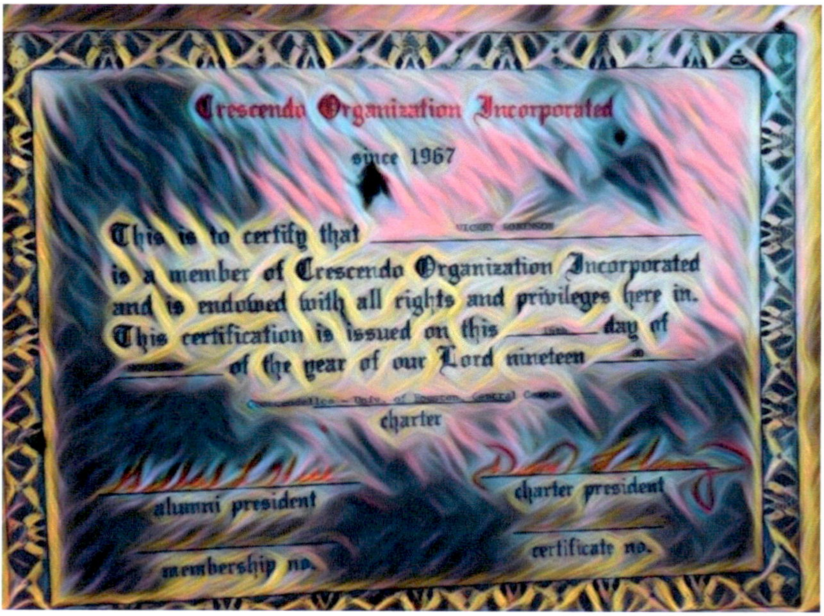

Crescendo Organization Incorporated

since 1967

This is to certify that ______
is a member of Crescendo Organization Incorporated
and is endowed with all rights and privileges here in.
This certification is issued on this ______ day of
______ of the year of our Lord nineteen ______

charter

alumni president

charter president

membership no.

certificate no.

I joined the Crescendo Organization, and with that organization, I found friendship and fellowship!!

Joining an organization transformed my time at the university.

I began to thrive socially; it truly enhanced my college experience.

And so, the journey continued!

My last few years at the university were not all smiles and laughter, but I made it through, received my degree, and was on my way to becoming a journalist.

I had my degree in hand, and it was time to officially pursue the dream!

I began my journalism career at a local newspaper as a correspondent journalist.

Having been promoted to staff writer, I was finally living the dream I had long cherished. I was on the path I had always admired, following in the footsteps of the journalistic idols I looked up to.

But wait…something in me had changed. The burning desire I had for years to be a journalist had morphed into something else.

I no longer wanted to continue the journalist path and was unsure why. But I did know that I had a new pressing desire: I wanted to join the military.

Why had this desire resurfaced? I couldn't say for sure, but the desire was strong, and I felt strongly that I had to act on it.

The Pursuit

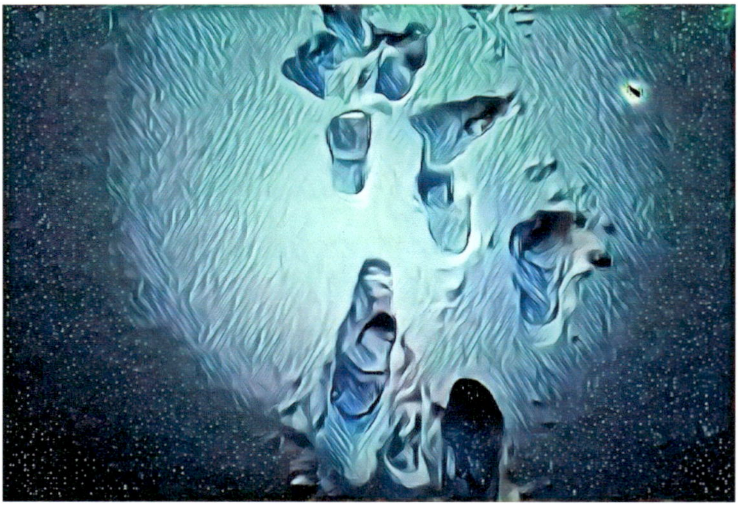

While working at the newspaper and completing my graduate degree in television journalism, I went and checked on my military entrance possibilities.

It was now 10 years later.

It was as if the sunlit trail was steering me toward something special. And it was shortly after venturing on the journey, I found myself at a military entrance processing station.

Something was tugging at me to follow the military path. With several needs gnawing at me – writing, traveling, leaving my comfort zone, pursuing adventure – the military was my way to take them all on simultaneously.

And although I was a decade older, I went for the military again. I took the entrance exam and physical and passed, again.

A few months after deciding to attempt the military again, the processes and paperwork for entry into the military were done. I handed in my notice at the newspaper, and as the time grew near to depart for training, a mix of fear, anxiety, and curiosity started to wash over me.

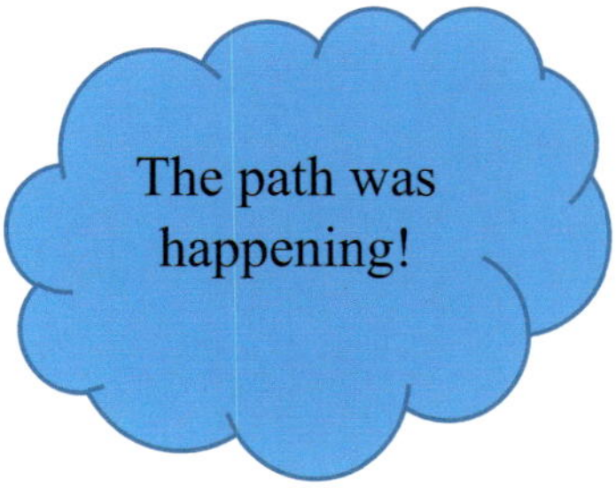

On the day I was to fly out for my new journey, I began questioning my decision. I was about to leave a promising career, a city I grew up in, and my family. I was about to leave everything I was familiar with for the complete unknown.

There were indications that there were to be no more spontaneous journeys on my part for a long time to come.

No more spontaneous trips to the beach.

The indications suggested that my time would no longer be mine to maneuver.

The path had led me to the doors of the military reception center.

Reception Center

Surprise!

I had arrived. From the moment I stepped off the plane to my arrival at the reception center, everything was meticulously organized and structured. There were coordinators at every turn, guiding my every move. I didn't need to worry about what came next—there was always someone there, ready to direct me and tell me exactly what to do next.

Day 2

My first full day at the center was structured with military precision. The sergeant meticulously outlined my schedule: where I needed to be, when to be there, what to wear, and even what and when to eat. The day was packed with activities—receiving vaccination shots, being issued uniforms, and learning the basics of military drills.

Surprisingly, the environment was calm and orderly, devoid of the yelling or hostility I had braced myself for. The staff's cordial demeanor was unexpected, making my transition smoother than anticipated. By late evening, I was briefed on my duties for the following day; I would be on guard duty from 1:00 to 3:00 am. Was this really happening? Yes, it was, and thus began my new path.

My new reality hit me at 12:30 am when I had to wake up and put on my full duty uniform, preparing for my guard duty that started at 1:00 am.

Day

3

No extended stay expected for the reception center, but I was not sure. Trainees were not privy to many details. We, the trainees, spent day three with a bit of physical training and cleaning of sleeping areas.

The path seemed as though it was about to get rocky. Tensions were seeming to rise!

Day 4

The day was stress-free. To begin, I got to sleep in; sleeping in meant getting up at 7:00 am as opposed to 4:00 am. It was nice. It felt like freedom.

There were many of us awaiting the beginning of official training and that meant many individuals with varying personalities trying to co-exist in the same space. With so many personalities, it led to many micro-conflicts.

Being an older trainee, I found myself in a different mindset compared to my fellow trainees. I had already made it through many of the thought processes they were venturing into. The mix of diverse personalities and experiences among the trainees created a challenging living environment.

Days into my new journey, the thing I missed the most about my old life was the peace and quiet.

Day

5

I suppose that when you have a large group gathering from all levels of society, you should expect some tension.

By the fifth day, our focus shifted to preparing for our departure from the reception center. The primary requirement to progress to basic training was to complete just ONE correct push-up; I achieved that right-of-passage. It was only then that I understood basic training would be the true beginning of my military journey.

Day 6

Day six marked my final day at the reception center. The next stop, the basic training station. Truly, after the initial stress of leaving all that was familiar, adjusting to the crazy hours, and relinquishing all control of personal time to someone else, the week at the reception center was doable. I wondered if the next phase of the military journey would be comparable to the initial phase. I had hoped so! And so, the journey continued.

Next Stop: Basic Training…

Made in the USA
Las Vegas, NV
07 November 2026

57669926R00019